LOVE OF THE LAND

LOVE OF THE LAND

JOEL AND LAVONNE STRASSER

Text by DORIS STENSLAND

Copyright © 1995
by ZIP FEED MILLS, INC.
Tom Batcheller and Joel Strasser

Produced by JOEL STRASSER PUBLISHING CO.
5109 W. 55th St.
Sioux Falls, SD 57106 605-362-0625

Photography and Concept by
Joel and Lavonne Strasser

Writing by Doris Stensland

Design and Typesetting by Terrie Miesen and Diane Martinson, Sioux Falls, SD

Printing Representative Everbest-Midwest, Edina, MN
Printed by Everbest Printing Co. Ltd., Hong Kong   First Printing 1995
No part of this book may be reproduced in any form whatsoever without permission
from the publisher.

Library of Congress Catalog Card Number: 95-092392

ISBN: 1-880552-03-5

Tom Batcheller
President
ZIP FEED MILLS, INC.

Those of us fortunate enough to spend our lives close to agriculture know of the true beauty that exists in every day rural life. The land, buildings, equipment, livestock, plus the other parts and parcels that together make up rural life, have provided both joy and sorrow throughout the years.

Joel Strasser has been capturing glimpses of rural life for over 45 years. His photographs grace the walls of our corporate office at ZIP Feed Mills, Inc. In 1987, to commemorate our 50th anniversary, ZIP Feed presented our dealers with framed prints of "Three Herefords" and "The Pig Looking through the Fence". These nostalgic photographs are prominently displayed in many of our dealerships.

ZIP Feed Mills commissioned LOVE OF THE LAND to commemorate our 60th anniversary. Please enjoy with me these views depicting life in rural America, as seen through the eyes and the lens of Joel Strasser.

Tom Batcheller

SOMETIMES we're so busy complaining about winter that we miss the little quiet joys, the "winter joys" that can give that warm happy feeling. What might some of these be?

A valentine morning, when the world is frosty and lacy; the sparkling whiteness of the snow when the sun shines on it; the security and shelter of home on a stormy night; round snow-topped corncribs that resemble frosted cupcakes; tracks in the new fallen snow.

And how about . . . fresh doughnuts, the aroma of baking bread, a good book on a wintry evening, a game of checkers . . . and popcorn? None of these give as much enjoyment in the heat of summer.

These days a little squirrel is busy scampering up and down the big cornpile that is located across the road from my kitchen window. This little fellow is having the time of his life. I suppose it is something like turning a child loose in the candy store – there is so much to pick from, he just doesn't know where to start.

When you live on the farm, you end up feeding many kinds of wild life. The pheasants, the deer that stop by, the squirrels, rabbits and coons . . . all help themselves. But you'll not hear the farmer complaining too much.

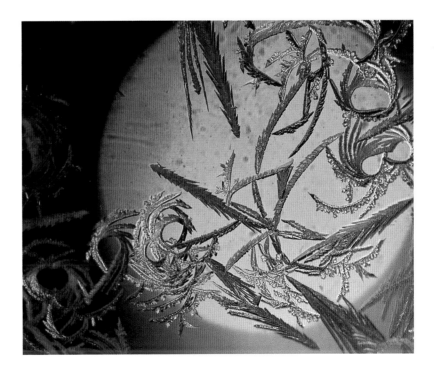

The geese flying northward usher in the season of Spring. As I watch them pass overhead, I often wonder if they dread this long, tiring journey every year . . . or do they merely consider it a game of "<u>follow the leader</u>."

MANY FARMERS are being real shepherds these days.

Now they don't just let their sheep wander in the field. At lambing time their charges need extra care. The ewes require special attention; the delicate little lambs need personalized care and often must be hand fed.

The little wild plum trees are in bloom. Mother Nature has placed these white bouquets here and there along the countryside, between the fields and by the fencelines. It's hard to pass one of these trees without stopping to pick a few branches.

IT IS ALWAYS GOOD to see some puddles again. These much appreciated and waited for raindrops are really pennies from heaven.

The gifts of Spring. Everywhere we look, we are treated to beauty: carpeted lawns of bright green; the brilliant colors of the tulips; apple trees in bloom; little lambs, calves and colts scampering about.

But what would Spring be like without two eyes to see it all?

It's Spring, and the farmer knows he has to get the seeds into the ground if he expects a harvest. And every day counts.

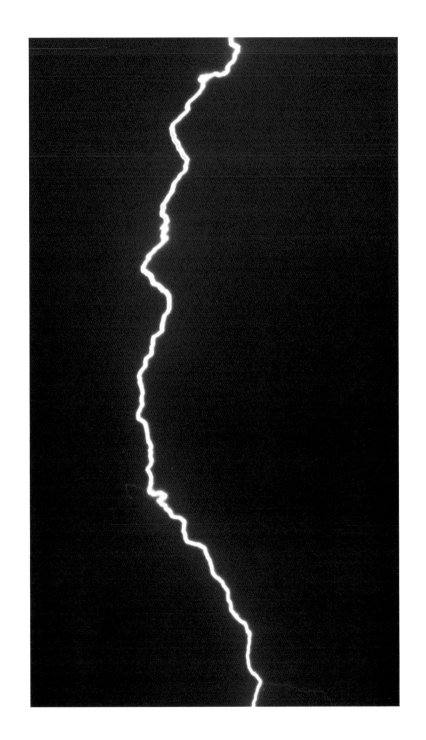

CLOUDS.

Fluffy white ones that look like puffs of smoke lazily floating by.

Wispy light gray clouds that resemble dust mice being chased across a blue sky floor.

From time to time we see angry dark ones piling up on the horizon. These give us an uneasy feeling and we wonder what they have in store.

The farmstead is the heart of the farming operation. Every day the farmer travels the treeless acres surrounding the farmstead, disking, planting, cultivating, and harvesting his crops. At noon and evening, he returns to the farmstead with its trees and flowers, which is an oasis where he is refreshed and gets rested for the next day's labor.

"Pigs will be pigs!"

Now they have joined our diet world. They are trying to maintain a long and lean look instead of the heavy, fat-laden hog image of yesterday, and want to be known as "the other white meat".

But the pig's heart hasn't changed. They still love a mud puddle, and they still possess their old selfish ways.

Memorial Day. For most of us it was a memory day. We came to these quiet places with our hands full of bouquets and our hearts full of memories . . . of a little child that didn't stay very long with us . . . of dear mothers, fathers or grandparents that had finished their tasks here on earth . . . of our other loved ones that had made our life richer for having known them . . . and also of the men who fought to keep our country free. These people are gone but not forgotten.

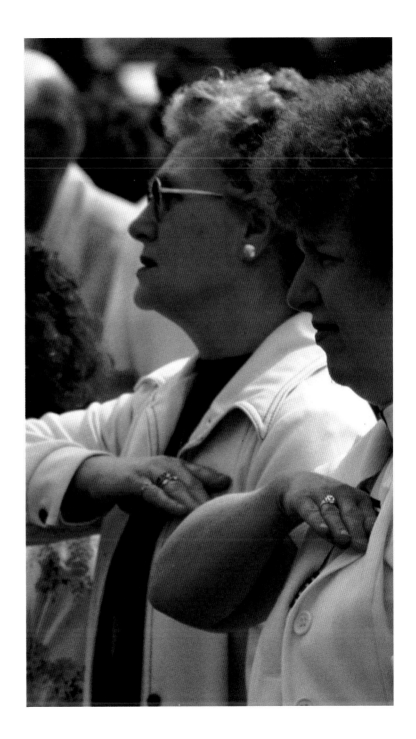

After getting a look at the fences that had been covered with snowbanks in winter, the farmer began to wonder if little gremlins had been busy doing mischief under the snow. It was hard to believe that plain snow could so completely destroy the well-groomed look they had last fall. Now any spare time the farmer finds will be needed for stretching and mending fences.

THERE ARE so many, many jobs to be done all at once that the farmer is cramming about two-days' work into every 24 hours. Daylight Savings Time allows him an after-supper workday. You will find most of the farmers going full speed from dawn's dark till evening darkness sets in.

This busy routine almost makes him feel like he is on a treadmill. After a day of chores, hours on the tractor seat or handling hundreds of bales, he reaches evening weary and ready for rest . . . and dreams of someone saying, "Take a week off!"

29

There is so much activity on the farm now that it is almost like a three-ring circus. In one field you will see a group of troopers of assorted sizes parading up and down the beanfield rows. In another field, muscular figures juggle hay bales, and in still another area you can watch a magic weed disappearing act as the farmer sprays his fields. Noisy wagons passing the house on their way from the field to the silo announce oat silo-filling just as certainly as the calliope tells it's circus time.

At threshing time these bundles were loaded. The hayracks filled with bundles stood in line like grocery carts waiting to be checked out at the supermarket. Then they were unloaded into the big threshing machine. Out came wagonloads of oats and big piles of shiny straw.

AT THE END of day, just before dusk, a hush comes over the land. The birds twitter their evening lullabies, the cattle leisurely graze on the hills and the cows relax after being milked and fed.

The sky takes on beautiful colors as the sun begins to set.

Then I am reminded of Millet's famous old painting, "The Angelus." . . . Two simple farm folks in their field at sunset, their heads bowed in prayer.

It is just as if all nature quietly folds her hands and waits for the evening benediction.

IT'S FASCINATING to watch the big combines work. These expensive pieces of equipment represent a tremendous investment.

The combine's giant appetite
isn't content with just a bite.
It licks the windrows off the ground,
Digests them with a growling sound.
When its hopper can hold no more
it spits out the grain on the truckbed floor.

Every farm has its own Water Department, and sooner or later it gets the problems that go with it. Sometimes it is even necessary to take on the project of digging a new well.

A farmer feels that things are going fine as long as his water supply keeps coming. Out of the earth's depths must flow this necessary liquid that fills our glasses, our bathtubs and our washing machines, quenches the thirst of our livestock and waters our crops. But getting it into our water pipes can sometimes be a costly and long-lasting endeavor. Usually most well-digging is done to the accompaniment of some moaning and groaning.

Time was when the wind blew and turned the windmill wheel that pumped the water for the farming operation. And there was a time when the little building behind the house was in daily use. Then electricity arrived! This caused both the windmill and the outhouse to come into disuse. Yet, each served a purpose in its day.

If we could take the railroad tracks back in time, we would discover a more leisurely paced life - the old blacksmith working at his own speed, the teams of horses slowly pulling up to a small elevator to unload their wagons full of grain. What a contrast to the hurried procession of trucks and semi-trucks arriving at large grain terminals today!

If you are driving along the highway you will notice that in the pastures the animal mothers are out strolling with their babies. Little calves are frisking alongside the stock cows. A flock of sheep has little lambs frolicking amongst them. Once in a while you may even discover a long legged new colt wobbling along with its mother. These little animal babies are very good at demonstrating their joy of being out by their fancy hops, skips and jumps.

Family farming is father, mother, brother, sister and God all joining together in the project of growing crops, animals, and people, experiencing love while doing it.

Family farms help make America GREAT. They not only produce food for the nation's tables, but they produce young men and women who make good Americans.

It doesn't take many minutes of visiting with a stranger before you discover if he has ever lived on a farm. If he asks intelligent questions and can converse for any length of time on farm subjects, you know he didn't spend all of his life in the city. Farm talk can be almost like speaking another language to someone who doesn't know about cornheads and moisture tests.

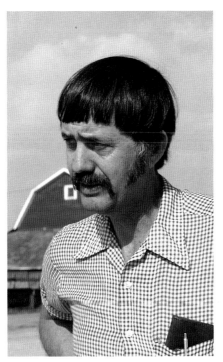

In its day the barn stood erect and tall, anxious to be of use, but time has taken its toll, and today the barn is old and tired and slouches against the landscape. Such is life! The time comes for wrinkles, bald heads, hearing aids and eyeglasses. We all need a laughing tree to remind us that a little humor makes it more bearable.

Ala
eme
soci
lini
Gra
egg

crat

These rural school houses hold many memories. Here several generations have learned reading, writing and arithmetic. They had their advantages and their disadvantages.

Outside many of these schools there still stands an old pump and outdoor-type restrooms as reminders of earlier days. The heating facilities of the past weren't modern either. Then the teacher had to get to school early in order to start the fire and have the room warm.

The cow is "a fountain of nourishment for the human race". Its four white streams each morning and night are the old man's sustenance and the baby's delight. Our menus would be very different if we didn't have the milk, butter, cheese and cream derived from the generous, faithful milk cow.

The process of obtaining the milk has changed from the farmer balanced on a one-legged stool, snuggled against the cow and milking "slow motion" by hand, to the very efficient milking parlors of today. But the milk remains the same healthful white drink it always has been.

The farmer takes pride in the make of equipment he uses. He'll argue long and strong on its merits. If he is a green machinery man you won't catch him with a piece of red equipment . . . and vice versa.

THE MILKWEED is busting out all over. Its silky insides are travelling hither and yon in search of a plot for next year's residence. Now their tall lanky skeletons are ready to be a part of someone's fall bouquet.

AND WHEN we're counting our blessings, let's not forget one of the best — "friends"! They are the golden threads that God weaves into the tapestry of our lives. They are placed along life's pathway so we have someone not only to share our times of joy and sorrow, but a friend is what the heart needs all the time.

To a Friend in Sorrow
    Dear friend . . .
    My heart, too aches because you grieve.
    Each thought begets a prayer for you.
    And though we may be miles apart,
    I weep today
    . . . with you.

MANY THINGS have changed through the years . . . but not September! She still behaves the same. And her goldenrod and browning cornfields still tell us that the year is two-thirds' spent . . . just as gray hairs remind us our lifetime is quickly passing.

Soon the goldenrod, bright flower of the fields, will be gone. And sometimes we must be reminded that our lives pass almost as quickly so that we spend our days wisely.

Each year when the corn husks get dry and their leaves start turning brown, the farmer makes a special trip into his cornfield to spy out the land — to see if the crop is rich or poor, and to determine if it is ready for harvest.

He walks up and down the rows, picking ears of corn at random, and then brings home a sample of the fruit of his land.

At a glance the cornfields in October look like wasteland, desolate and unkempt. But the farmer knows there is treasure hiding there. During the summer these cornstalks have been green factories quietly manufacturing their product. Beneath these dry husks hang ears of gold. The farmer doesn't hoard these nuggets, but hauls them to market so the whole world can be nourished.

This "love of the land" is a mighty force that keeps pushing a man through thick and thin. He will spend his whole life wooing these black acres of loam.

He will gamble his savings to feed her. He will borrow to keep her. He will scrimp and save so he can afford her. But it will be worth it all if in the fall she emerges bountiful and beautiful.

In the fall, nature's wild fowl become gun-shy. During hunting season their colorful garb and feathered apparel make them a prime target. The only shooting that is not fatal to them is done by a camera.

What's for dinner?

In bible times, the fatted calf was killed and roasted as a special treat for guests. Beef is still a very important item on the menu, and raising, feeding, selling and processing it is a large industry today.

The items up for auction may represent
many years of farming, but there is one thing
that did not go up for auction. It is the
memories of the years they spent on the farm,
of the dreams that did come true, of a little
family growing up here, of the years of
working, loving and living. These are treasures
that can never be taken away from them.

Snow.

Sometimes it falls on the earth softly and silently. At
other times, each snowflake arrives with stinging fury. When
it teams up with the wind, the whole out-of-doors appears to
be throwing a tantrum.

Spring, summer, fall and winter – time does not stand still. Life has its seasons also. Spring brings youthful dreams, beauty, energy and potential.

Summer is a time of fruitfulness and accomplishments.

In fall, life slows down, nests are empty and there is time to enjoy what we have made of our lives, or what life has made of us.

The winter silhouette stands stark, lonely and unadorned, showing off scars, broken boughs and broken dreams. But it can also reveal the beauty of a character that has been shaped by faith and love.

A special thanks to Tom Batcheller, President, ZIP FEED MILLS, for making this book, *Love of the Land* possible . . . for Doris Stensland, her writing, that has such a feeling for rural life . . . the tremendous privilege we've had creating the visual and the concepts for this book . . . the closeness to nature has reminded us of the writing in Isaiah . . . THE GRASS WITHERETH, THE FLOWER FADETH: BECAUSE THE SPIRIT OF THE LORD BLOWETH UPON IT: . . . BUT THE WORD OF OUR GOD SHALL STAND FOREVER.

Joel and Lavonne Strasser